say it with

Candy grams

MOUNDS
of thanks
for all you
DEW

GO
STATE!

CANDIES NOT PICTURED ELSEWHERE IN THIS BOOK THAT
MAKE GREAT ADDITIONS TO CANDYGRAMS:

Printed in China

Products

507 Industrial Street
Waverly, IA 50677

ISBN-13: 978-1-56383-449-3
ISBN-10: 1-56383-449-9
Item #2909

candygram (kăn'dē grăm) n. – Candy delivered with a greeting or other prepared message.

Say it with Candy

WHAT BETTER WAY TO MAKE SOMEONE'S DAY THAN A GIFT OF FAVORITE CANDIES AND A FUN MESSAGE? HERE'S THE INFORMATION YOU NEED TO MAKE CANDYGRAMS EXTRA SWEET.

You'll find poems, phrases and tags for just about any gift-giving occasion, but candy bars often stand in for some of the words to make each message fun. Simply attach the corresponding candies as listed so the recipient can read the entire candygram. Whether assembling a poster, bouquet, card or care basket, make each gift unique by mixing and matching the phrases to suit the occasion and personality of the recipient.

Use the photos and follow the brief directions for basic assembly, but choose the format you like best. Make the tags any way you like and use lettering that matches the theme. Here are some additional tips for successful messaging...

Poster Pointers

* Use poster board, foam core board or cardboard display boards for a sturdy base. Sheets may be purchased in white, black and other colors in various sizes and shapes.

* Poster bases may also be cut from large cardboard boxes. Fasten poster board on top with glue dots.

* To trim foam core or cardboard, use a utility knife against a yardstick on a cutting mat.

* Outline posters with wide masking tape, colored duct tape, markers, paper strips or stickers to "frame" them, or try hot-gluing beads or other trims around foam core board.

Containers

* Lidded jars and bottles may be any shape, but those with a wide mouth work best for pieces of wrapped or unwrapped candies. Buy them new or use recycled ones.

* Place Stryofoam in containers when making bouquets and cover it with tissue paper, excelsior moss or shredded paper.

* Plastic sandwich bags can be folded and taped to the correct size for any project.

* Paper lunch sacks and gift bags are just the right size for most candygram care packages or survival kits. If desired, cut holes in the front and line the inside with cellophane or plastic wrap to make windows.

Lettering Tips

* Water-based or permanent markers show up well on white posters, but try paint markers to print words on colored posters.

* Use paper, vinyl or foam alphabet stickers, or cut out paper letters and glue them to the base.

* Try using rubber stamps with ink pads for both lettering and decorations.

* Write the messages in straight lines or clustered around the candy bars. Arrange candy on the base and lightly print your message in pencil to decide what works best for your space and candy.

* Since candy bar names are standing in for words in the message, cover up unnecessary letters (or words) with stickers. In some cases, you may need to add letters with permanent markers or make arrows pointing to the word you want recipients to read. The candygram texts and photos will guide you.

4

Candy Connections

* Candy bars come in a variety of sizes so choose the size that best fits your project, from king-size to miniature.

* Lightweight candies may be attached to a candygram with double-sided tape, craft glue or a glue stick.

* Attach heavy candies to the base with glue dots or a low-temperature glue gun. (Hot glue will melt some candies.)

* If candygrams will be displayed upright, be sure the candies are firmly attached.

Tags & Labels

* Design tags on the computer and print them with colored ink. Cut and mount on heavier paper as needed.

* Use cardstock or scrapbook paper for tags and signs. Layer different papers together or use mounting tape to add interest and dimension.

* Write messages on purchased tags, such as key tags or punch-outs.

* Attach tags with ribbons, cords, jute, brads and other items. Reinforce hanging holes with tape or grommets if necessary.

HAVE FUN WHEN YOU CREATE AND DELIVER THESE SWEET TREATS - YOU'LL MAKE SOMEONE'S DAY!

$\mathcal{S}$AY IT WITH SODA

Gather single bottles or cans of soda (or a whole 6-pack) and say:

* My "**POP**" is **A&W**esome! Happy Fathers' Day
* You're an **A&W**esome friend!
* We're **ROOTIN'** for you! Go team!
* Thanks for **POPPIN'** in

* **MOUNDS** of thanks for all you **DEW**

* I'm **SODA**lighted

* Good luck – I'm sure you'll **DEW** well!

MOUNDS
of thanks
for all you
DEW

A SIMPLE POUCH MADE FROM HEAVY PAPER HOLDS THIS MOUNDS CANDY BAR.

What would I **DEW** without you?

7-UPlifting ways to start the school year

1. Wake Up with a smile.
2. Get Up and get moving.
3. Clean Up with a hot shower.
4. Charge Up with a good breakfast.
5. Listen Up when the teacher talks.
6. Speak Up when it's time to share your thoughts and knowledge.
7. Reach Up to achieve your goals.

* **CRUSH** the competition!

* I **SODA** think you're amazing!

* I have a **CRUSH** on you!

* Love Potion #9

7

BIRTHDAY POSTERS

TOP 10 THINGS TO DO FOR YOUR 30th

10 Ignore the [TIC TAC] of the clock
9 [TAKE 5] and do something [NUTRAGEOUS]
8 Go on a shopping [SPREE] along [5 AVENUE]
7 [ROLO]ver and get [GOOD & PLENTY] of rest
6 Enjoy a '[PAYDAY]' with [100 GRAND]
5 [LAFFY] a lot – or at least [SNICKER] a little
4 Go to a [Symphony] or view an [eclipse]
3 Share your [HUGS], [KISSES], and [NUGGETS] of wisdom
2 Have [MOUNDS] of fun doing [WHATCHAMACALLIT]
1 Go to [XTREMES] to have a good time!

TOP 10 THINGS TO DO FOR YOUR 30TH

10 Ignore the **TIC TAC** of the clock

9 **TAKE 5** and do something **NUTRAGEOUS**

8 Go on a shopping **SPREE** along **5TH AVENUE**

7 **ROLOVER** and get **GOOD & PLENTY** of rest

6 Enjoy a **PAYDAY** with **100 GRAND**

5 **LAFFY** Taffy a lot – or at least **SNICKER**s a little

4 Go to a **SYMPHONY** or view an **ECLIPSE**

3 Share your **HUGS**, **KISSES** and **NUGGETS** of wisdom

2 Have **MOUNDS** of fun doing **WHATCHAMACALLIT**

1 Go to **XTREMES** to have a good time!

8

TO OUR CHILD

It's great to have a **FUN** ~~Dips~~ kid like you...
At that age between **RUNT**s and **BIG LEAGUE CHEW**.
The **S'MORE**s I watch you learn and grow
The **S'MORE**s I have to let you know
You're a **SWEETART**s, but I can't lie –
I'm **SHOCKE**rs**D** at how the years fly by.
I'm **EXTREME**s**LY** proud of all you do
Happy Birthday, **SUGAR BABY** – I love you!

OVER THE HILL

So you're 50. Don't get your **MILK DUDS** in a bunch.
Sure, your mid-section might get **CHUNKY**, but
there's no **RIESEN** to worry unless the **ROLOVER**
your pants interferes with tying your shoes. If you
get **BUTTERFINGERS**, just get a grip and **TAKE 5**.
And if freckled **DOTS** and wrinkles appear and your
brain turns to **SKITTLES** and you call the couch a
WHATCHAMACALLIT, we'll just give you a **HUG** and
say **YOR**k a **TREASURE**s. Now, go be a **SLO POKE** –
you've earned it!

9

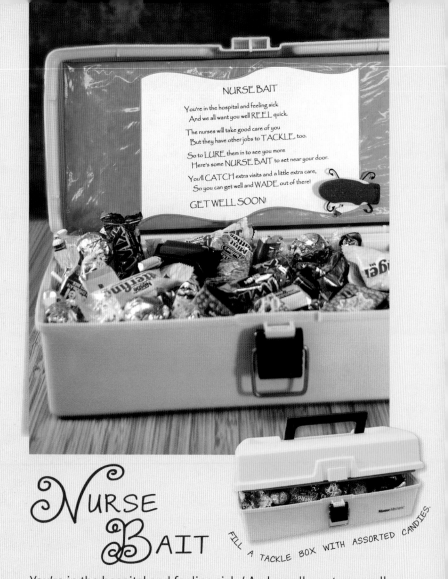

Nurse Bait

FILL A TACKLE BOX WITH ASSORTED CANDIES.

You're in the hospital and feeling sick / And we all want you well **REEL** quick. / The nurses will take good care of you / But they have other jobs to **TACKLE**, too. / So to **LURE** them in to see you more / Here's some **NURSE BAIT** to set near your door. / You'll **CATCH** extra visits and a little extra care, / So you can get well and **WADE** out of there! / **GET WELL SOON**!

CHRISTMAS

EXAMPLE USES
24 CARDSTOCK
SQUARES AND THE
CIRCLES INSIDE
THE MINI MUFFIN
PAN SPELL OUT
"HAVE A VERY
MERRY CHRISTMAS."

Spell out a message on 1″ paper circles and tape them to the inside of mini muffin cups. Add small candies. Cover openings with numbered paper shapes, attaching them with tape or magnets.

OTHER HOLIDAYS

EASTER: Fill numbered plastic eggs with Easter treats and place them in a basket. Pick an egg each day.

JULY 4TH: Place numbered stickers on plastic snack containers and fill with mini candies wrapped in red, white, blue and silver. Open one container each day.

HALLOWEEN: Cover **TOOTSIE POPS** with white tissues and gather with white twist ties to look like ghosts. Draw on eyes and a mouth and write a number on each one. Eat a ghost every day until Halloween.

I really SKORed with a
FUN or friend like you!
EXTREME, CHUCKLES &
SNICKERS in everything we do.
You're my LUCKY CHARM,
when we go arm in arm
When it's left UP2U, it's fun!
WOO-HOO!
Happy Birthday!

ℋappy ℬirthday

Print the following poem on a large paper flower and attach to a craft stick. Place in a basket and fill with matching candies.

I really **SKOR**ed with a **FUN** ~~Dip~~ friend like you / **XTREME**s **CHUCKLES** & **SNICKERS** in everything we do / You're my **LUCKY CHARM**s when we go arm in arm / If it's left **UP2U**, it's fun! Woo-hoo! / Happy Birthday!

Fill a glass with candy; top with a Styrofoam ball and insert sucker sticks and straws. Add shredded paper and a tag.

* 30 **SUCKS**! But you're sweet!

* My life would **SUCK** without you.

* You **BLOW** ~~Pop~~ me away!

LAYER SCRAPBOOK PAPER ON SUCKERS TO MAKE FANCY FLOWERS.

Cut flower shapes from decorative paper, making a hole in the center. Cover **SWEETARTS** rolls with green paper for stems. Slide flowers onto stems and push stems into Styrofoam in a container. Add excelsior moss, a few paper leaves and a label.

(TRY OTHER CANDY STEMS LIKE ZONKERS, SHOCKERS, BOTTLECAPS OR MENTOS.)

15

$\mathcal{L}$OVE POSTERS

EXAMPLE USES POSTER BOARD AND WHITE PAINT MARKER.

To my **SWEETART**s,

You make my ~~Chocolate~~ **HEART GUSH**ers – can't you see? / Your **HUGS & KISSES** are enough for me! / I count my **LUCKY CHARMS** to have this **HOT TAMALE**s in my arms. / I'm ~~Salted~~ **NUTS** ~~Roll~~ about you, **SUGAR DADDY**! / **ZOTZ** & **ZOTZ** of love – **YOR**k**S** truly, (name)

HAPPY ANNIVERSARY

To my **HUBBA BUBBA**,

It was **MINT** to be when we married [25] years ago! Each year you steal another ~~Reese's~~ **PIECE**s of my **HEART**. You're **MOUNDS** of **FUN** ~~Dip~~ and we share plenty of **SNICKERS** each day. I don't need a shopping **SPREE** or a diamond ~~Candy~~ **NECKLACE** or to do the **MAMBA** – it's the **SIMPLE PLEASURES** that I **TREASURE**s. Expect some **RED HOT**s **HUGS** and **KISSES** to help celebrate this **MOMENTOS** day! Happy Anniversary, **SUGAR DADDY**! [or ~~Bit-O-HONEY~~]

GENERAL LOVE

5 RIESENS you make my **HEART** skip

1. You **TREASURE**s our life together
2. I like the way your **HONEY BUNS** look in jeans
3. You're the best **SUGAR DADDY**
4. If I get a little **CHUNKY**, you don't notice
5. You always give me a **BIT-O-HONEY**

NOW & LATER and forever, I'll love you!

Or say...

* Your **HUGS & KISSES** send me into **ORBIT**
* You bring **GOOD & PLENTY** of **BLISS** into my life every day
* We are **MINT** to be together
* **U-NO** that I love you, **NOW & LATER** and forever

You're a big sister! Oh the fun you will **SKOR**! A sweet little **SUGAR BABIE**s just for you to adore! You'll go fishing for **GOLDFISH**, play games the whole ~~Pay~~**DAY**, search for **TREASURE**s like pirates, find **STARBURSTS** & the **MILKY WAY**. Watching you play will make Mom's ~~Chocolate~~ **HEART** flutter, and if you are lucky, you'll get a **NUTTER BUTTER**!

BABY GIRL

Sweet as an angel, soft as a **DOVE**, / A **BIG HUNK** of my heart is ready to love! / She's a little **BIT-O-HONEY**, with **SPICE DROPS** thrown in, / She'll smile and make **BUBBL**icious**S** and your heart she'll win. / Dressed all in **LAYERS** and cute as a bug / She's a **RIESEN** to celebrate – and perfect to hug!

BABY BOY

2 **CHUNKY** legs and 10 tiny toes / Your precious **BABY** ~~Ruth~~ boy is a ~~Almond~~ **JOY** to behold. / **HUGS & KISSES** to share, **MOMENTOS** galore / From **ROLOVERS** and **SNICKERS**, **GOOD & PLENTY'S** in store. / His **CHUCKLES** of mischief, you'll take all in **STRIDE**. / "I'm **NUTS** about you, (child's name)," you'll say with great pride.

TWINS

Oh **BABY** ~~Ruth~~... Oh **BABY** ~~Ruth~~...
DOUBLE~~mint~~ the **FUN** ~~Dip~~...
You have **RIESENS** to celebrate
With two, not just one!!

On tag: You're Dynamite!!!

WRAP 'EM UP

Wrap **ROLO** rolls in red paper and fasten three rolls together with black electrical tape. Use glue dots to attach string to the top and add a tag to say:

* You're **DYNAMITE**!!! or You're the **BOMB**!

* **BOOM, KA-BOOM, ZOWIE, POW!** You're the one that's got it all!

* **BOMBED** the test? Please don't fret. Eat some candy and take a rest.

Put **ORANGE SLICE CANDIES** in a windowed bakery bag and add a label to say:

* **ORANGE** you glad to be a (team name)?

* No matter how you **SLICE** it, you're a winner.

Wrap a **TAKE 5** candy bar in a paper sleeve and add a label that says:

* **TAKE 5** – You've earned it!

* **TAKE 5** – Lights, Camera, Action!

Cut and crease green cardstock like a matchbook to enclose a **MINT GHIRARDELLI SQUARE**. Staple the candy wrapper inside the bottom of book and add a label on the front saying:

* Happy St. Patrick's Day

* We're a **MATCH** made in heaven!

* You **STRIKE** my fancy!

Attach a heart-shaped tag to a bag of **CHOCOLATE HEARTS** and say:

* You're always close to my **HEART**

* **HEART**-throb!

* I ♥ you

* It's **HEARTENING** to know I've got your support. Thanks.

Attach a holiday label to a pack of **TWIZZLERS** that says:

* **TWIZ** the season to be jolly!

* You send my heart into a **TWIZZLE**!

* I get all **TWIZZLY** when I think of you

Attach a tag to a **100 GRAND** bar and say:

* I wouldn't trade our time together even for **100 GRAND**. I'm the luckiest girl in the land!

* You're ~~100~~ **GRAND**!

Tape a **CRUNCH** bar to a large cardstock rectangle and add a message that says:

* In a **CRUNCH**, you're always there for me. Thank You!

* After **CRUNCHING** the numbers, I've found that you're #1!

Fill a jar with **BUBBLE GUM** and add circle labels that say:

* Thanks for **CHEW**SING to be a Teacher – I had a **BALL** this year!

* May your day **BUBBLE** over with fun!

* It takes plenty of **GUM**PTION to do what you did!

Attach a starburst-shaped sign and ribbon to a colored craft stick and insert into one end of a **STARBURST** box to say:

* A new **STAR** has **BURST** onto the scene. Good job!

* In case you need a **BURST** of energy

FOLD CARDSTOCK INTO A CARD (ABOUT 5½" x 8½"). GLUE ON CONVERSATION HEARTS AS PART OF THE TEXT (CANDIES ARE NOT EDIBLE).

HEY BABE—

The **BEST DAY** ever was when you said, "Please **BE MINE. MARRY ME!**" All you had to do was **ASK ME**. You knew I'd **SMILE** and **SAY YES**. You always know how to make me **LAUGH** and **BE HAPPY**, and your **SWEET TALK** makes me **LOVE YOU**. You are my **ANGEL** and I will always **B TRU 2 U**. You're such a **CUTIE PIE** and **UR HOT** too! [**WINK WINK**] I love to **HOLD HANDS**. And when you **HUG ME**, you show me such **SWEET LOVE**. Remember our **FIRST KISS**? I gave you a **HIGH FIVE**. I knew right then it was more than **PUPPY LOVE** and you were my **LOVE BUG. U R A STAR** and I am **CRAZY 4 U**. We know how to **DREAM BIG** and make **TIME 2 DANCE**. I know you are my **SOUL MATE. U R IT** for me, **BABY DOLL**. I just want to tell you this Valentine's Day that you are still the **LOML** [love of my life]. It's **TRUE LOVE**! **XOXO** (name)

24

Our pop rocks!

We think our **POP ROCKS** (Inside: Not to **BUTTER**finger you
up, but it's surely no **SHOCK**ers we think our ~~Sugar~~ **DADDY**
is worth **100 GRAND**! / With love from your **3 MUSKETEERS**
[or **RUNTS** or little **GOOBERS**]

Use colored duct tape to attach
two pieces of foam core board
(9 x 13″) to form a book. Attach
candy with glue dots.

25

DOOR HANGERS

Create a cardstock pouch to hold large rolls of **SMARTIES**. Add flower cut-outs and a tag that says, "Congrats **SMARTIE**!" Tape a ribbon hanger to the back.

Or do this: Fill a plastic bag with small **SMARTIES** and staple a folded cardstock sign over the top to say, "Good Luck, **SMARTIE** pants. You'll do great!" Add a jute hanger.

26

Fill a plastic bag with homemade **"PUPPY CHOW"** (or purchase **MUDDY BUDDIES**). Add a ribbon and tag to say:

* My **DOG-GONE** big wish: May there always be **PUPPY CHOW** in your dish.

* Hope you'll feel better in two shakes of a **TAIL**.

* **"PAWS"** and relax with a treat!

* Life's **"RUFF"** without you; hope to see you soon!

* You lucky **DOG**! Congratulations!

Tie a ribbon loop through the hanging hole in a **STAR BRITES** package and attach a label with this nursery rhyme:

Star light, star bright,

First star I see tonight,

I wish I may

I wish I might

Have the wish

I wish tonight!

May all your wishes come true!

27

ꙅRANDPARENT POSTERS

Grandpa,

A **SOUR PATCH** ~~Kids~~ may seem unfair, / but in a **CRUNCH**, you're always there. / Sharing **TREASURES** of your past / life's **MOMENTOS** that will last. / **CHUCKLES & SNICKERS** - I'm so glad / for **GOOD & PLENTY** times we've had. / **NOW & LATER** & forever more, / these **SIMPLE PLEASURES** I'll adore. / You're the best!

GRANDMA

Grandma,
YORKIND and sweet and oh-so **NICE**, / Offering **HUGS** without thinking twice. / **NOW & LATER**, great times we'll share. / I'm **OVER**~~Almond~~**JOYED** by how you've cared. / Your little grand**GOOBER**s

GRANDMA & GRANDPA

MR&MRS (last name), but Grandma & Grandpa to me / You're the best **DUO** on our family tree. / Love you to ~~Reese's~~ **PIECES**, **NUTS** about you / **KUDOS** for everything that you do. / This is to tell you I think you are ~~100~~ **GRAND** / And I am the luckiest kid in the land. / **ZOTZ** and **DOTS** of **HUGS & KISSES**, / (name)

GRANDMA & GRANDPA

Sweeter than a **BIT-O-HONEY**
WERTH~~ers~~ more than tons of money
~~100~~ **GRANDMA**, ~~100~~ **GRANDPA**, you're the best
ZOTZ more fun than all the rest!

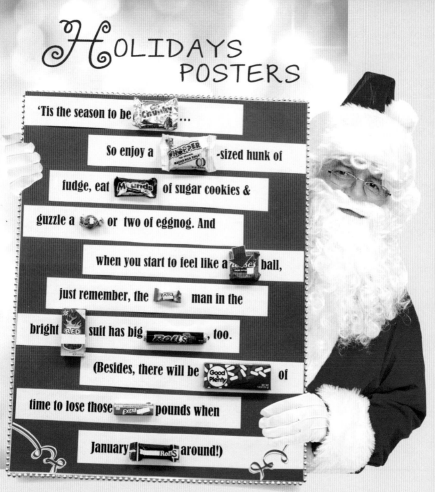

'TIS THE SEASON

'Tis the season to be **CHUNKY**... So enjoy a **WHOPPER**s-sized hunk of fudge, eat **MOUNDS** of sugar cookies & guzzle a **NIP**s or two of eggnog. And when you start to feel like a ~~Reese's~~ **PEANUT BUTTER** ~~Cup~~ ball, just remember the **JOLLY** ~~Rancher~~ man in the bright **RED** ~~Hots~~ suit has big **ROL**o**LS**, too. (Besides, there will be **GOOD & PLENTY** of time to lose those **EXTRA** pounds when January ~~Tootsie~~ **ROLLS** around!)

HALLOWEEN FUN

If you hear a little **KᵣACKLE**,
Then a **SNICKER**s or two,
Be ready to make a **FAST BREAK** –
Don't turn to **LOOK** behind you!
Ghosts, goblins and **TWIX** abound,
Trying to scare you to ~~Reese's~~ **PIECES** – BOO!
But be a **SMARTIE**s – don't be afraid
Just **CHUCKLE**s at them and they'll run from you!

ST. PATRICK'S DAY

A **FAMOUS** ~~Amos~~ bit of Ireland
I'm sending to you today
To bring you **LAYERS** of happiness
On this fine St. Patrick's ~~Pay~~**DAY**
It's a wee and tiny present
Full of ~~Mr.~~ **GOOD**~~bar~~ luck and many wishes
A little four-leaf clover
And lots of Leprechaun **KISSES**

SO GLAD YOU'RE MY SISTER!

Trident Layers Gum
Sometimes we have to sort through the layers to
find the good stuff that holds us together.

Tic-Tacs
Because the clock keeps ticking, we need to
remember to live our best life every day.

Twizzlers
We might be a little twisted,
but we like each other that way.

Animal Crackers
There are a lot of animals out there.
We just need to bite the bad ones.

Chocolate Coins
Carry a little chocolate and a little
spare change wherever you go.

Smarties
Sisters teach each other a lot...
thank you for passing along your pearls of wisdom.

Pez
For dispensing joy into my life.

Now & Later
I love you now & I'll love you later. Always

FOLD, LAYER AND PUNCH PAPER CUPCAKE LINERS
OVER BAG TOP AND INSERT PICK.

FOR SIS

Fill bags with candies listed in poems. Attach or insert the message.

SO GLAD YOU'RE MY SISTER!

TRIDENT LAYERS GUM ~ Sometimes we have to sort through the layers to find the good stuff that holds us together. / **TIC TACS** ~ Because the clock keeps ticking, we need to remember to live our best life every day. / **TWIZZLERS** ~ We might be a little twisted, but we like each other that way. / **ANIMAL CRACKERS** ~ There are a lot of animals out there. We just need to bite the bad ones. / **CHOCOLATE COINS** ~ Carry a little chocolate and a little spare change wherever you go. / **SMARTIES** ~ Sisters teach each other a lot... thank you for passing along your pearls of wisdom. / **PEZ** ~ For dispensing joy into my life. / **NOW & LATER** ~ I love you now & I'll love you later. Always and always!

VOLUNTEER SURVIVAL KIT

Volunteer to use your many talents – it's all **UP2U**. / Opt to help often – we are in need of **GOOD & PLENTY** of enthusiastic volunteers. / Laugh a little (**SNICKERs-SNICKERs**). / Utter these words: "I **CHEWS** to stay positive." / Cut loose and be a little **NUTTY** now and then. / **TREASUREs** the rewards that come from helping others. / **EXTRA** hands make light work. / Energize yourself with a **FAST BREAK**. / RE~Almond~**JOYCE** in helping others. / Savor the satisfaction of a job well done. **KUDOS**!

STAMP A BROWN BAG AND ADD A CARDSTOCK AND RIBBON TOPPER.

It's gonna be SWEET

Countdown to Vacation

SUN	MON	TUE	WED	THU	FRI
28	27	26	25	24	23
21	20	19	18	17	16
14	13	12		10	9
7	6	5	4	3	2

PILL BOX ATTACHED
TO FOAM CORE BASE

COUNTDOWN CALENDARS

Cover each "door" of a plastic pill box with a numbered sticker. Mount the box on a larger foam core board base and decorate the top. Fill each opening with small candies. Open one door each day to find a treat and count the days until vacation.

7 Days to an Extra Ice Sweet Birthday

SUN

MON

TUES

WED

THURS

FRI

SAT

BIRTHDAY

Open an envelope and moisten the flap's adhesive. Place the front side of a second envelope over the flap, lining up its bottom edge with the flap's fold. Press to hold. Attach five additional envelopes in the same way. Tape a poster board arrow with hanging ribbon to the flap of the top envelope and say, "7 Days to an **EXTRA** Sweet Birthday." Write the days of the week on the envelopes and fill with candy. Empty one envelope each day until countdown is complete.

OR TRY THIS...

Make a paper chain from sturdy paper and attach a piece of wrapped candy inside each numbered link. Hang the chain up and tear off one link each day until school starts – or ends.

Put wrapped treats inside numbered balloons. Blow them up and tie them closed. Tack the balloons to a bulletin board. Pop one balloon each day to get a treat and count the days to a birthday or another special event.

ℬLAST ℴFF

Cut a rocket from glitter paper and outline it with **STARBURSTS**. Attach **PIXY STIX** for the flames and say, "It's Your Birthday! Have a **BLAST**!"

Or say:

* **BLAST** off to a new year

* You're the **STAR** in my universe

Arrange **TWIX** and **YORK** candies on poster board to spell "**I LOVE YOU**." Outline candy with markers before attaching.

Add these phrases in heart shapes:

Just **BETWIXT** you & me

YORKIND of lovin' warms my heart

Arrange mini **DOTS** boxes in a flower shape on poster board and roll up **ZOTZ** strips in the center. Add the message, stem and leaves with markers. Fasten green **ZOTZ** candies on leaves, if you like. Say: You add **ZOTZ** & **DOTS** of **BLOOMIN'** fun to my world!

FOR TAG, USE COLORED INK TO MATCH JELLY BEAN COLOR.

Jelly Beans for my teacher...
pink because your smile glows
green because you helped me grow
white is for your understanding
orange means you're so outstanding
yellow for your joy each day
black for helping pave my way
purple for your service yearly
red because you care sincerely
This jar of jelly beans just for you
to say I'm thankful for all you do!

THANK YOU TEACHER!

JELLY BEANS for my teacher...

PINK because your smile glows / **GREEN** because you helped me grow / **WHITE** is for your understanding / **ORANGE** means you're so outstanding / **YELLOW** for your joy each day / **BLACK** for helping pave my way / **PURPLE** for your service yearly / **RED** because you care sincerely / This jar of **JELLY BEANS** just for you / to say I'm thankful for all you do!

RAINY DAY CHEER

> If all of the raindrops were *lemon drops* and *gumdrops*, oh what a rain that would be.
>
> **Wishing you a sweet rainy day!**

Fill plastic bags with **LEMON DROPS** and **GUM DROPS**. Cut two pieces of heavy paper (6˝ x 8˝ and 6˝ x 10½˝). Crease 1¼˝ from one end of each piece; crease a 2½˝ flap on other end of long piece (bag back). Cut out raindrop shapes on smaller piece (bag front). Overlap and staple the 1¼˝ flaps together (bag bottom). Tape candy bags behind raindrops; fold flap over bag front, punch holes and insert ribbon. Add a tag to say, "If all of the rain drops were **LEMON DROPS** and **GUMDROPS**, oh what a rain that would be. Wishing you a sweet rainy day!"

A THIRST-QUENCHER

Thanks for quenching my thirst for knowledge

Fill a glass with **GRAPE KOOL-AID PACKETS** and say:

* Thanks for quenching my **THIRST** for knowledge

* Thanks for sparking my creative **JUICES** this year!

* You're **KOOL**!

* To one of the **KOOLEST** kids I know!

ℭONGRATS POSTERS

Dear _____,

As you ~~Tootsie~~ **ROLL** toward graduation, you can be **EXTRA** proud. You've crammed **MOUNDS** of facts into your brain & spent **SKORS** of late nights studying so your grades are **GOOD & PLENTY**. You are such a **SMARTIEs** – not an **AIRHEADs** or **DUM DUM**! **NOW & LATER**, you've taken a **FAST BREAK** for some **SNICKERS** or a shopping **SPREE**, but **KUDOS** for always **CHEWSING** to stick with it. You will hear a **SYMPHONY** of cheers as you march by in your ~~Bottle~~**CAPs** and gown. You ~~Pop~~ **ROCKs**! Congratulations! We are proud of you. Enjoy your **MO**MENTOS accomplishments! ~~Chocolate~~ **HEART**, Your family

GENERAL CONGRATS

Congrats on your **WHOPPER**s of an accomplishment! **EXTRA** hours, **MOUNDS** of hard work and ~~Tootsie~~ **ROLLING** with the ~~Sour~~ **PUNCHES** ~~Straws~~ is now **SKORING** you a big **PAYDAY**. So, **TREASURE**s this **MOMINT**, then keep aiming higher; the **MILKY WAY** is your limit!

COLLEGE BOUND

Soon you'll be one of the college **SMARTIES**
So please invite me to all your parties!
I'd like to meet:

The **CRACKER JACKS** who make you grin,

The **HOT TAMALES** that lure you in

The **AIRHEADS** that are sort of "dim"

The **CHUNKY** kid they call **SLIM JIM**,

The rich kids who have **GOOD & PLENTY**

(Who in a **CRUNCH** could loan me a '20').

Please introduce me to **MARY JANE**,

To **MIKE AND IKE** and what's-her-name,

To the **SUGAR BABY** with eyes of blue

Plus the **DOUBLEMINT** twins and **KITKAT** too

KUDOS to you, friend – college life is a **TREASURE**s, / You'll have **MOUNDS** of fun – simply too much to measure.

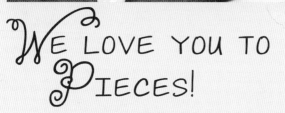

WE LOVE YOU TO PIECES!

Fill a jar with **REESE'S PIECES** and attach a tag to say:

* The **PIECES** all fit when we work together. Thanks for everything.

* I'd go to **PIECES** without you!

* Honey, you stole a **PIECE** of my heart.

* Thanks for sharing a **PIECE** of your mind.

Layer foil-covered chocolate coins and **SKITTLES** in rainbow order and say:

* Happy St. Paddy's Day!

* **SKITTLE** over the rainbow to find your pot of gold!

* A little sweet luck to you!

* You're the pot of gold at the end of my rainbow.

Fill with **PEANUT M&M's** and say:

* For my **NUTTY** Dad

* SWEET on the outside, **NUTTY** on the inside – just like YOU!

* I'm **NUTS** about you, Sweetie!

* You may be a little **NUTTY**, but you're still sweet to me.

Use **CANDY EGGS**, wrapped or unwrapped, to say:

* For my **EGG**cellent teacher

* Great news! We're **EGGS**pecting!

* Your help has been **EGGS**ceptional.

* No more **EGGS**cuses. I'll do better next time.

FILL & GIVE

The M&Ms of Menopause:

Chocolate cravings? Eat the **BROWN** ones.
Yes, ALL the M&Ms are chocolate, but show some restraint!

Hot flashes? Reach for the **RED** ones.
and just keep eating until they're gone.

Forgetfulness? Eat the ORANGE ones.
They probably won't help you remember, but they will sure make you feel better.

Moody? Go for the "calming" GREEN ones.
If it works, call me. If not, please don't call.

Depressed? Eat the **BLUE** ones.
When you eat the last one, smile.

Loss of bladder control? Yup, grab the YELLOW ones.

Weight gain? Well, of course!!
You've been eating lots of **M&Ms!**

The **M&M's** of Menopause

Chocolate cravings? Eat the **BROWN** ones. Yes, ALL the **M&M's** are chocolate, but show some restraint! / Hot flashes? Reach for the **RED** ones, and just keep eating until they're gone. / Forgetfulness? Eat the ORANGE ones. They probably won't help you remember, but they will sure make you feel better. / Moody? Go for the "calming" GREEN ones. If it works, call me. If not, please don't call. / Depressed? Eat the **BLUE** ones. When you eat the last one, smile. / Loss of bladder control? Yup, grab the YELLOW ones. / Weight gain? Well, of course!! You've been eating lots of **M&M's!**

Fill a candy jar with **assorted candies** and say:

* Grandpa's **'STACHE**

* A Sweet **'STACHE** for your Sweet Tooth

* Your own private **'STACHE**

* Just for you from my special **'STACHE**

USE A WOOD DOWEL "FISHING POLE" AND PAPER "HOOK" LABEL.

Put **GUMMY WORMS** in a colorful bucket to say:

* **HOOKED** on You

* I'm **HOOKED**

* Thanks for watching my sq-**WORMY** kids!

* A treat for when the kids are being sq-**WORMY**!

Fill a glass with **RIESEN** candies and say:

* 21 **RIESENS** to Celebrate Your Birthday

* Need a **RIESEN** to smile? Here are lots of 'em!

EXAMPLE USES A BEER GLASS WITH 21 RIESENS.

GREATEST POP

Fill a container with pop-related items such as soda, popcorn (microwave and packaged snacks), pop-themed candies, **POP TARTS** and lollipops like **BLOW POPS** or **TOOTSIE POPS** and add a tag that says, "For the World's Greatest **POP**!"

Or say:

* Please **POP** over for a visit soon!

* **POP** Quiz: Who's my favorite niece?

* You're a **POP** star in my world!

$\mathcal{P}$ICK ME!

Fill a basket with apple-related items such as cereal bars, apple candies and gum, snack pies, whole apples and packages of apple cider drink mix. Add a tag to say, "I couldn't have **PICKED** a better teacher!"

Or say:

* Thanks for **PICKING** up the slack!

* Of all the apples in the orchard, I **PICKED** the best – YOU!

* An apple a day keeps the doctor away – get well soon.

49

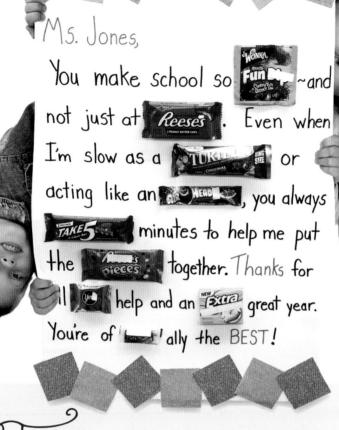

THANK YOU POSTERS

(Teacher's name),

You make school so **FUN** ~~Dip~~ – and not just at **REESE'S**. Even when I'm slow as a **TURTLE** or acting like an **AIRHEAD**s, you always **TAKE 5** minutes to help me put the ~~Reese's~~ **PIECES** together. Thanks for all **YOR**k help and an **EXTRA** great year. You're ~~Swedish~~ **OFFISHALLY** the best!

YOU'RE GREAT!

Your **MOJO'S** a **SHOCKER**s / You're **S'MORE**s than just great / With the **EXTRA** hours you put in / To me, you're first rate!

THANKS DAD!

Dad,

We considered sending you on a ~~Swedish~~ **FISHING** trip or **SKORING** you tickets to a **BIG RED** Sox game or even a new pet ~~Kit~~**KAT**, but decided we'd rather ~~Breath~~**SAVER**s our time with you. You're a real ~~Werther's~~ **ORIGINAL** and we **TREASURE**s our ~~Sugar~~ **DADDY** days.

Love, your little **RUNTS** and **RED HOT**s mama

[Or say: **HUGS & KISSES**, Your little **GOOBERS**]

TEAMWORK

I can't thank you enough / Your work was **XTREME**s / We make a great **DUO** / Glad you're part of my team!

51

Party Favors

Place **COOKIES** in zippered baggies and staple a folded cardstock sign over the top to say:

* Sugar & Spice & Everything Nice (**SNICKER DOODLE** cookies)

* Chip Off the Old Block (**CHOCOLATE CHIP** cookies)

* Everything's sweeter with a little chocolate! (**CHOCOLATE** cookies)

* Rx: Milk & Cookies ~ Feel Better Soon! (any type)

With ribbon, tie a plastic spoon and key tag label to a small box of **CEREAL** and add a label that says:

* ⁎ I **CEREALSLY** like you (any cereal)

* ⁎ **CEREALSLY**, I'm loopy about you (**FRUIT LOOPS**)

* ⁎ I'm **CEREALSLY** sweet on you (any sweet cereal)

Roll paper and secure ends to create a cone. Attach cellophane to inside lip of cone, fill with **TOOTSIE ROLLS** and close. Add a tag to say:

* ⁎ You're sweet, kind and fun... all **rolled** into one!

* ⁎ You're on a **ROLL**! Keep up the good work.

Put a bag of **TRAIL MIX** into a blue jean pocket and add a tag to say:

* ⁎ Happy **TRAILS** to you

* ⁎ Blaze a **TRAIL** – you've got what it takes!

* ⁎ Hike on over to see us sometime!

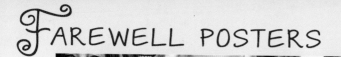

You've made it! Retirement's gonna be ~~100~~ **GRAND**! / Bring the **BAZOOKA** – strike up the band! / That big **CHUNK**y of money you **SKORED** on **PAYDAY** / Can be used for fun and not ~~Tootsie ROLL~~ away. / The **NICE** folks you worked with will be missed/ But days without alarm clocks will be **BLISS**. / So make a **FAST BREAK** before they ask you to stay / **STRIDE** away quickly – it's your time to play!/ **KUDOS** to you!

GOING AWAY

While we **CERT**s**AINLY** don't make **100 GRAND** working here, we've had **GOOD & PLENTY** of laughs, which is **WERTH**er's Original far more. I'll try not to fall to ~~Reese's~~ **PIECES**, but will miss your **KINDNESS**, ~~Tic~~ **TACKY** jokes, and silly **TWIX**. Good luck to you **NOW & LATER**!

GOOD-BYE

Good luck – may you find a **STARBURST** of ~~Almond~~ **JOY**

Observe – new surroundings offer **MOUNDS** of opportunities

Open the door to **FUN** ~~Dip~~ new experiences

Don't be a stranger – **TAKE 5** to drop everything and say "hi" to your new neighbors

Be happy – There's a **BIT-O-HONEY** in every new situation

Yesterday is behind you – **GOOD & PLENTY** of things lie ahead

Embark – Take a deep breath and have a **NUTRAGEOUS** time!

We'll miss you to ~~Reese's~~ **PIECES**!

EXAMPLE USES COLORED POSTER BOARD ATTACHED TO FOAM CORE BOARD.

(HUBBA BUBBA GUM)

ℱEEL BETTER

Arrange a box of tissues, bottle of honey, cough drops, socks, soup, oranges and Hugs and Kisses candies in a container with shredded paper. Fasten the sign to a paint stir stick and insert in basket to say:

Tissues to blow your **BIG RED BAZOOKA** / A **BIT-O-HONEY** to soothe your sore throat **(OUCH!)** / **GOOD & PLENTY** of cough drops for your nagging cough / **LAYERS** of socks to help keep you warm / **CHUNKY** soup to warm you up from the inside / **JUICY FRUIT** for **EXTRA** Vitamin C / Lots of **HUGS & KISSES** to let you know I care!

ℬEE-ATTITUDES

Bee-attitudes to help your new grandchild GROW!

Place the verse in an envelope and matching candies in a container.

BLISS ~ Be prepared for the feeling in your heart that's pure BLISS.

EXTRA gum ~ Be grateful for this EXTRA-precious baby.

HUGS & KISSES ~ Be generous with your HUGS & KISSES.

CHOCOLATE HEART ~ Be warm and say "I love you" often.

NUGGETS ~ Be willing to pass along your NUGGETS of wisdom.

KIND bar ~ Be KIND – your words mean the world to him/her.

SUGAR BABY ~ Be gentle – your BABY needs a soft touch.

ALMOND JOY ~ Be willing to share your JOY with him/her.

MOUNDS ~ Be dependable and make MOUNDS of memories together.

TREASURES ~ Be humble – TREASURE Mom & Dad's opinion and remember what it was like to be in their shoes.

SWEETARTS ~ Be patient because every now and then, even the SWEETest child can be a little TART.

REESE'S PIECES ~ Be open. Share PIECES of your story so s/he will know about your family.

SIMPLE PLEASURES ~ Be carefree and teach your grandchild to love life's SIMPLE PLEASURES.

We always have a **WHOPPER** of a good time together.

You & I were **MINT** to be friends.

You make life a ... e **JOY!**

I like every little **PIECE** of you!

MALLOW out with me.

You're a **LIFESAVER** to me.

ℱRIENDSHIP IS 𝒮WEET

Choose the items you want in the basket and create tags for each one using any of these phrases.

BARTON'S MILLION DOLLAR BAR ~ Your friendship is worth a MILLION.

SNICKERS ~ Your SNICKERS make me smile.

SMARTIES ~ You're such a SMARTIE pants – but like it!

POP ROCKS ~ A friend like you really ROCKS!

TOOTSIE ROLL or SALTED NUT ROLL ~
I like how you ROLL.

NERDS or NERDS ROPE ~ We may be NERDS sometimes, but we have lots of fun together.

JUNIOR MINTS ~ You and I were MINT to be friends.

WHOPPERS ~ We always have a WHOPPER of a good time together.

REESE'S PIECES (or Almond Joy Pieces) ~ I like every little PIECE of you!

ALMOND JOY ~ You make life a pure JOY!

MALLOW CUPS ~ MALLOW out with me.

CARAMELLO or MELLO YELLOW soda ~ MELLO out!

LIFESAVERS ~ You're a LIFESAVER to me.

GOLD COINS ~ Don't ever CHANGE.

CHUNKY ~ Feelin' funky? Try a CHUNKY!

ZERO BAR ~ ZERO hour is upon us, but we can do it!

WERTHERS ~ Under the WERTHER? I can help.

EXTRA GUM ~ You always go that EXTRA mile!

INDEX

Birthday
Baskets & Bouquets 14-15
Countdown Calendars 35
Jars & Other Containers 47, 56
Posters 8-9, 36

Congratulations
Door Hangers 26, 27
Posters 42-43, 54-55

Encouragement / Thinking of You
Door Hangers 27
Individual Candies 20, 21, 23
Jars & Other Containers 39, 41, 46, 47
Party Favors 52-53
Posters 37
Say It with Soda 6-7

Family
Baskets & Bouquets 48, 59
Cards 25
Countdown Calendars 34
Jars & Other Containers 44-45, 47
Posters 9, 18-19, 28-29, 51
Say It with Soda 6,7
Survival Kits 32-33

Friendship
Baskets & Bouquets 15, 60-61
Care Packages 41
Door Hangers 27

Party Favors.................................... 52-53

Individual Candies 20-23

Jars & Other Containers 44-45, 46-47

Say It with Soda............................... 6,7

Get Well

Baskets & Bouquets 10, 49, 58

Door Hangers................................... 27

Holidays / Special Events

Countdown Calendars

Any special event.......................... 35

Christmas 13

Easter 13

Halloween...................................... 13

July 4 13

Vacation, School........................... 34, 35

Valentine's Day............................. 12

Individual Candies

Christmas 22

St. Patrick's Day............................ 21, 22

Posters

Christmas 30

Halloween...................................... 31

St. Patrick's Day............................ 31

Jars & Other Containers

Announcing Pregnancy.................. 45

Father's Day.................................. 45

St. Patrick's Day............................ 45

Love

Cards 24
Jars & Other Containers 44, 45, 47
Countdown Calendars 12
Individual Candies 20, 21, 22
Posters 16-17, 37
Say It with Soda 6-7

New Baby

Baskets 40
Posters 18-19

Retirement/Going Away

Posters 54-55
Door Hangers 27

Students

Care Packages 11, 41, 57
Individual Candies 20-23
Posters 42-43

Teachers

Jars & Other Containers 38-39, 45
Baskets & Bouquets 49
Posters 50

Thank You

Jars & Other Containers ... 38, 39, 44, 45, 47, 57
Individual Candies 23
Posters 28-29, 50-51
Say It with Soda 6-7

Volunteers

Survival Kits 33
Jars & Other Containers 57